GEORGE DON
THE FORFAR BOTANIST

"man of genius"

Edward Luscombe

The Pinkfoot Press
Brechin
2007

Published 2007 in Scotland by

The Pinkfoot Press, 1 Pearse Street, Brechin, Angus DD9 6JR

In collaboration with

Angus Council Cultural Services

ISBN 9781874012535

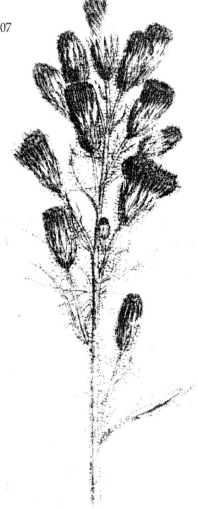

1 *Alpine sow thistle*

Cover: Glen Clova (photograph © William H B Richards)

Printed by Robertson Printers, Forfar

Contents

List of Illustrations

Foreword

I am glad to commend this remarkable small work. Dr Luscombe has undertaken a great deal of research into the life of an extraordinary son of Angus who deservedly merits this fitting tribute.

The book is easily read and gives a fascinating insight into one man's love for, and dedication to the beauty which he discovered in The Botany of Angus-shire (as Don called his County); Glen Clova in particular had an immense attraction for him. He did all this at great personal cost. His own life ended in poverty, but it had greatly enriched the scientific knowledge of his own day and of subsequent generations.

We owe a debt of gratitude to Ted Luscombe for reminding us of the life and work of this forgotten genius. George Don richly deserves to be widely read by anyone interested in our native botany and in the history of our Royal Burgh.

Bill Middleton
Provost of Angus
April 2007

2 *George Don's vasculum (specimen collecting box) in the Meffan Institute*

By Way of Explanation

This monograph had its origin in a conversation in 2006 with the Provost of Angus (Councillor Bill Middleton) and his wife (Councillor Glennis Middleton). I had just completed a book on Episcopacy in Forfar since the Reformation when they suggested that I might try to do something to rescue from obscurity the name of the eighteenth-century botanist George Don, Forfar's forgotten genius.

A switch from ecclesiastical history to botany (even in a modest way like this) involves a steep learning curve. I owe a real debt of gratitude to the Middletons and to the following for helping me in that learning process and in the preparation of this little book:

The Staff of Angus Archives, Restenneth

Mr James Carnegy-Arbuthnott, Balnamoon

The Staff at Forfar Loch Country Park

The Staff at Forfar Public Library

The Family of the late Duncan Fraser, Montrose

Canon Dendle French, Glamis

Andrew N. Gagg's Photo Flora

Mrs Shelagh Millar, Easter Ogil

Dr Henry Noltie, the Royal Botanic Gardens, Edinburgh

Mr William H. B. Richards' Photography

The Staff of the Meffan Institute, Forfar

Edward Luscombe
Kirkton of Tealing, Angus
2007

3 *Portrait of George Don Snr.*

I   A Man of Genius

In the introduction to her book on Scottish plant hunters (*Seeds of Blood and Beauty*, Ann Lindsay, 2005) the author refers to the enormous contribution made by Scottish botanists to the horticultural scene around the world.

> A handful of Scots, roughly the same in number as a smallish class or the Scottish football or rugby teams with a few reserves on the side, have hunted out and introduced into the West more plants from around the world than, say, all the other European nations combined. Every garden in Britain, and in most of our European neighbours as well, contains plants originally brought back to Europe by a Scot. From the mid-sixteenth century onwards Scots' expertise, cunning, curiosity, intelligence, adventurousness of spirit, scientific knowledge and business acumen dominated the horticultural scene, not only in the United Kingdom but around the world … Yet theirs remains an untold story.

For its size the County of Angus produced a surprisingly large proportion of that 'handful of Scots'. In the early years of the nineteenth century when the population of Angus was only seventy-three thousand (less than half the size of present-day Dundee) six names in particular stood out. There were three members of a Forfar family – George Don and his two sons, George and David from Dovehillock; there were the brothers Thomas and James Drummond, sons of the Head Gardener on the Fothringham Estate at Inverarity; and there was Robert Brown, son of the Episcopalian minister at Montrose.

All of these achieved considerable eminence in their chosen scientific field. All except one were considered to be of sufficient merit for inclusion in the Dictionary of National Biography. The one exception is the subject of this monograph – George Don, Senior. His omission from the Dictionary is a surprising one, since he was in many ways, the foremost botanist of his day.

The six were not the only botanists from Angus during the nineteenth century. Charles Lyell of Kinnordy (1769–1849) had a special interest in roses and was the author of the (unpublished) *Flora of Kirriemuir*. Walter McNicoll (1827–1908), Land Steward at Tealing, shared with Lyell a passionate interest in geology as well as in botany. Almost contemporary with McNicoll was the Revd. John Stevenson (1836–1907), Parish Minister of Fern and author of both *Mycologia Scotica* and of *British Fungi*.

These three, together with George Don, Senior, and his son George, confined themselves principally to the flora of Britain. The Drummond brothers, David Don and Robert Brown were all intrepid world-travellers in the hunt for new species of plants.

'A man of Genius'. Those words appear on an impressive memorial obelisk that stands near the door of the Parish Church in Forfar's East High Street. They refer to George Don whose name is universally recognised and respected by botanists throughout Scotland and beyond. By contrast the name of this son of the Royal Burgh is scarcely known amongst the general public at the present day.

In 1877 the Police Commissioners of Forfar renamed a road in his honour. Dovehillock (or more commonly, Doohillock) Road became Don Street, an event commemorated by a less than obtrusive plaque on the wall of the Forfar Community Resource Store in the same thoroughfare. Yet the common belief is that the name refers to that other Don family – the textile manufacturers whose enterprise in the eighteenth century brought lasting economic growth and employment to the town, and at the same time prosperity to themselves. The contrast with George Don could hardly be more striking. Whilst he undoubtedly brought a great wealth of botanical discovery and knowledge to the world of natural history, he himself died in penury.

This is not a botanical treatise. It is simply an attempt to recover and make more widely known something of the man himself and of his contribution to scientific knowledge.

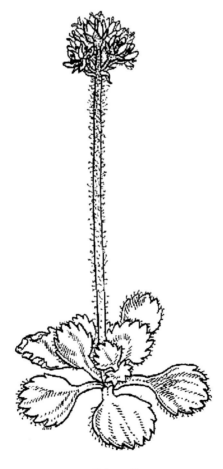

4 *Alpine saxifrage*

II The Early Years

George Don was born in the parish of Menmuir in Angus or Angus-shire as he insisted upon calling the county. Menmuir is a district of almost sixteen square miles lying to the west of Brechin and to the north of Careston. At the time of his birth, the population numbered some eight hundred souls. The sole occupation was agriculture. The largest estate lying to the east of the parish was, and has remained, in the ownership of the Carnegy-Arbuthnott family. For the rest, the smaller properties were then a mixture of owner-occupied and tenanted farms. It was on one of these, named Ireland, that Don was born in 1764. Ireland was on the Carnegy-Arbuthnott estate of Balnamoon. It was situated about two miles north of Menmuir in a cold back-lying position. There is no public road to all that now remains of the farm buildings – a pile of stones accessible only by a 4x4 vehicle.

It was a time of considerable change, as much in the county as in the rest of Scotland. Barely eighteen years had elapsed since James Carnegy, the Laird of Balnamoon ("the rebel Laird") had been forced to hide for six months in the hills above Glenesk because of his support for Prince Charles Edward Stuart in the '45 Rising. The place where he hid is still known as Balnamoon's Cave. His family were staunch Episcopalians and they and their fellow Episcopalians in Menmuir were in the pastoral care of the Incumbent of Tarfside in Glenesk. When George Don was born, the Incumbent was the Revd. James Brown whose son Robert was, like Don, to become one of the leading botanists of his day and who was later to be associated with Don in plant discovery.

The actual date of Don's birth is not known. What is certain is that he was baptised on 11 October, 1764 by the Revd George Ogilvy, the Parish Minister of Menmuir from 1750 until 1779. It is almost certain, too, that the baptism would have taken place in the home and not in the Parish Church. This was the usual practice at the time and for long afterwards. Even a century later no less than thirty-eight years were to pass without a single baptism being performed in the parish church.

The parish register records the father's name as Alexander Don. The mother's name was Isobel Fairweather. Both parents were descended from respectable farmers in the area, the Dons having originally come from Aberdeenshire and settled in Edzell in the seventeenth century. Alexander's occupation was given as a currier, a description generally assumed to be that of a shoemaker, although it could equally well refer to that of a groom. The family continued to live in Menmuir for a few years before moving for a short time to the neighbouring parish of Careston.

In 1772 or 1773 when George was eight or nine years old the family moved to Forfar. They took up residence in Little Causeway, a square off the West High Street deriving its name from the causeys or cobblestones with which the area was paved. Alexander had an uncle who worked as a shoemaker in the burgh, and he followed his uncle into the same trade.

5 *Little Causeway, Forfar in the nineteenth century*

It is impossible to say what formal education George Don received, but it appears likely that this was confined to the parish school in Forfar where reading, writing and arithmetic made up the elementary curriculum. Whatever the deficiencies may have been, he wrote in a bold well-formed hand and in a clear and vigorous style. He was described as being a boy with a natural talent for reading and observation – aptitudes which were soon apparent from his botanical discoveries and recording.

Whatever defects his formal schooling may have had, George Don's real education came from his powers of observation and his intense interest in natural history. He shared with his father a love of horticulture and whilst still at school regularly went for long rambles in the countryside around Forfar, exhibiting as avid an interest in fauna as he did in flora.

Whilst still a schoolboy he was reputed on one occasion to have returned to his native parish of Menmuir to stay with relatives. In the course of his visit he collected an assortment of wild flowers which he planted in the garden and classified in a system of his own devising.

III   A Wandering Gardener, He

Don's time at the Parish School in Forfar could not have been of any great duration. Local tradition had it that on leaving school he was sent to learn his father's trade of shoemaking, but soon discovered that his heart was not in boots and shoes.

It was not long before he was sent to Dunblane to become a watch and clock maker. This is borne out by the Forfar Parish Register which describes his occupation as 'watchmaker and botanist'. In Dunblane his interest in botany continued as great as ever. Dr. Patrick Neill, a regular correspondent of Don's, wrote 'He was first apprenticed to the clockmaker in the town of Dunblane, and here formed his first *hortus siccus* comprising all the mosses and flowering plants which he could cull from the neighbourhood, and they were numerous.' (A *hortus siccus* – literally a dry garden – is a collection of preserved plants, known to botanists as an herbarium). It was the first of a succession of similar collections that he was to make, and which contributed significantly to the knowledge of plants, native to Scotland in general, and to the County of Angus in particular.

Watchmaking in Dunblane obviously did not hold the young Don any longer than shoemaking had done in Forfar. In 1779, at the age of fifteen he was sent to Dupplin Castle, the seat of the Earl of Kinnoull. There he was placed under the supervision of the head gardener of Dupplin Gardens, who was married to Don's aunt, the sister of his father. It is impossible to say how long he remained there, but it was whilst he was in his first year there that he made his first botanic discovery. The discovery was of a new species of moss growing on a rock in Dupplin Den, and it was named after him – *Gymnostonum Doniana*.

In the same year he made the first of what were to become regular visits to the Scottish Highlands in search of plants. It was while he was at Dupplin that he first met the young woman who was later to become his wife. 'On one of his botanical rambles' wrote his grandson, 'he met a woman who was carrying on her head a pitcher of water. Entering into conversation with her and helping her by taking her burden from her, he began his acquaintance with Caroline Stewart, his future wife, an active energetic woman, as my father has described her to me'. Caroline was in service with the Oliphants of Gask.

Don used all his spare time at Dupplin to explore the Ochils and the Grampians, steadily increasing his knowledge of Scottish flora. From Dupplin he moved south to England. There he appears to have worked as a short-term gardener in a variety of places. Amongst others he spent some time in London, Oxford, Bristol, Yorkshire and Warwickshire. In all he was away from Scotland for almost eight years, although he did make frequent return visits to Forfar.

In 1788 he returned to Scotland and moved to Glasgow to resume his trade of watchmaking, but now as a journeyman. Once again all his spare time was spent in the search for plants. In 1791, he met up with Robert Brown, whose father had moved from Tarfside in Glenesk to become the Episcopal Minister at Montrose.

Robert Brown was also a keen botanist, but unlike Don he had followed an academic course of study, firstly at Marischal College, Aberdeen and then at the University of Edinburgh. At the time of their meeting in Glasgow Brown was an eighteen-year-old medical student. Together they went on a two week expedition to Angus, during the course of which they visited the Moss of Restenneth where they discovered a specimen of rush (*Scirpus hudsoniensis*) in its only known British location. In the following year Brown read a paper before the Edinburgh Natural History Society. This was an account of some of the rarest plants that he and Don had discovered during their tour of Angus. The paper was described as 'a model of its kind and an astonishing production for a youth of eighteen'.

Robert Brown made return visits to Forfar in each of the next three years, before going to Ireland as an assistant army surgeon. He was appointed naturalist to the expedition under Captain Matthew Flinders for the survey of the then unknown coasts of Australia and spent the years between 1801 and 1805 in gathering some four thousand plants, many of which were new. On his return to Britain he was appointed librarian of the Linnean Society. He went on to become one of the most noted botanists of his time.

Brown was not Don's only acquaintance during his time in Glasgow. He struck up a close friendship with another young botanist named John Mackay. Mackay had trained at Dickson and Company's Garden in Leith Walk, Edinburgh where one of Don's own sons was later sent as an apprentice. Don himself and Mackay made numerous extensive field-work expeditions which continued after Don had moved back to Forfar. Amongst other areas they visited was Ben Lawers when in 1793 they discovered Mountain Stitchwort (*Minuartia Rubella*). They spent some time in Glen Tilt where the Duke of Athol and his sister, Lady Charlotte Murray, had invited the two friends to botanise. In 1794 they separately paid a visit to the Isle of Skye. In 1800 Mackay was made Principal Gardener of the Royal Botanic Gardens in Edinburgh. It was a short-lived appointment for two years later he died at the early age of twenty-seven. There is a moving story of how Don tried to keep his friend's spirits up by spreading mosses over the table in an effort to revive his interest in botany.

6 *Alpine sandwort*

IV   Settling Down

In 1797 George Don obtained a lease of ninety-nine years on a plot of land on the north side of Forfar. By this time he had married Caroline Stewart, the young woman whom he had met and helped on one of his forays from Dupplin Gardens. Caroline was to give birth to fifteen children, although only six of them survived their father – five boys and one girl. All five boys were trained as gardeners and the two eldest sons, George and David, went on to become distinguished botanists in their own right. The sister died shortly after her father's death, and the grandfather, Alexander Don was dead within a fortnight at the age of ninety-six.

The plot of land was a feu from the estate of Charles Gray Esquire of Carse who let it to Don for a peppercorn rent, with a proviso that he should build a cottage dwelling-house on the site. The place was known as Dove Hillock (or more colloquially, Doo Hillock) from the small hill in the centre of the plot. Fortunately we have a clear description of the two acres and of the way in which it was laid out.

> The ground as a whole sloped to the west, and was shaped somewhat like a horseshoe. From all sides except the west it fell towards the centre from which a small oval-shaped knoll rose, known as the Dove Hillock. On this Don built his house. From the knoll the ground sloped down to the west into what had at one time been the bed of Forfar Loch. Here he formed a large artificial pond which he stocked with aquatic plants and fish. A wall ran round the garden, leaving room for a broad border, in which the native plants were arranged according to their orders in their appropriate soils, of which there was an abundance of all kinds at hand – loam, clay, sand, gravel and moss. In addition he rented several acres of land from the town, which were used as a nursery for young trees.

It was apparent that Don was much more concerned with substance than with appearance. Dr Neill commented that he was 'far from artistic in his presentation of the plants, and the garden made very little external show, needing well laid-out or well-kept gravel walks, hedges or edgings and a certain amount of ornamentation.' Part of the reason for this may well have been the fact that Doohillock also served as a market garden to sell produce in Forfar that would help to keep the growing Don family in food and clothing. Partly, too, because Don was frequently away on his plant-gathering expeditions.

The Doohillock Garden attracted numerous visitors, many of them nationally known botanists. The name of one of these, Dr Patrick Neill, has already been mentioned. He was an Edinburgh publisher who devoted himself to botany and horticulture. He was well known as the publisher of *The Flower, Fruit and Kitchen Garden* amongst other works. Neill described his first meeting with George Don

> When on a pedestrian excursion along the East Coast of Scotland, I happened to spend a night in Montrose, and it occurred to me that both Brechin and Forfar should be visited – the former for its well-known Den Noran and its round tower of remote antiquity; and the

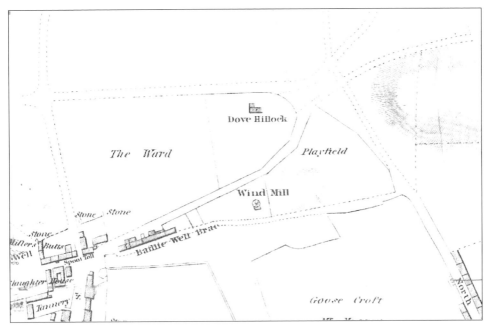

7 *Site of Dove Hillock, 1822*

latter for the remarkable botanic garden and its owner, whose fame was familiar to me owing to my intimacy with his regular correspondent, Mr. James Mackay of the Leith Walk Nurseries.

On reaching Forfar towards evening I soon found Don's garden and on entering enquired of a very rough looking person with a spade in his hand whom I took for a workman, whether Mr. Don was at home. The answer was "Why sir, I am all that you will get for him". Having apologised in the best way that I could, I stated that when I left home I did not anticipate a visit to Forfar, else I could have brought with me a note of introduction from Mr. John Mackay. Mr. Don, pointing to my botanical box, immediately said "That is introduction enough for me" ... I persuaded him to accompany me to the Forfar Inn where he spent the evening with me. Next morning by six he met me by appointment and conducted me to Restenneth Moss, where I had the great satisfaction of procuring a living patch of *Eriophorum Alpinum* (a species of the sedge family) which grew on the drier and firmer parts of the Moss. Mr. Don remarked that in a few years the plant would disappear – which I understand has happened!

A decade later, writing in the June 1809 edition of the *Scots Magazine*, Dr Neill commented that the flower garden and flower nursery at Forfar was scarcely surpassed in Britain in terms of number, diversity and rarity of hardy plants.

Mr. George Don has surmounted many difficulties in following out his favourite pursuit, and in forming so extensive and curious a collection of living plants. The whole of the plants are of a hardy sort, Mr. Don not possessing either greenhouse or stove for the protection of such as are tender. It is in alpine plants and hardy perennials and annuals the Forfar garden excels … Mr Don has introduced several hundred species of hardy plants, most of which have never before been cultivated in Scotland.

Almost from the beginning at Doohillock, George Don had carried on correspondence with a large number of Scottish botanists, whom he began to supply with specimens of plants that he had gathered on the hills, firstly from around Forfar and then from much further afield. This was in addition to the vegetables and fruit that he was selling to the residents of Forfar.

Dr Neill was by no means the only noted botanist to travel to Forfar. One visitor was the Right Reverend Dr Samuel Goodenough, the Bishop of Carlisle. On his arrival in Forfar, the bishop asked to be taken to see Mr Don. Not surprisingly he was led to the senior member of that other Don family mentioned in Chapter One – Colonel Don. "That is not the man" said the bishop, and was reluctantly taken by his guide to Doohillock to see the fellow-botanist whom he had travelled more than a hundred miles to meet. The bishop evidently found Don to be an impatient correspondent, expecting responses to his letters and queries by return.

The death of John Mackay had left a vacancy at the Royal Botanic Garden. The Garden had begun in 1670 beside the Palace of Holyrood House in Edinburgh, originally in an area hardly larger than a tennis court. It was described then as a 'Physic Garden' introduced primarily to grow plants used for medicinal or healing purposes. By the beginning of the nineteenth century it was starting to acquire a name as a leading centre in Scotland for botanical research and cultivation. In 1802 it was still located in its original site, not moving to the present location in Inverleith until 1821.

The vacancy left by Mackay was variously described as Principal Gardener or Superintendent of the Royal Botanic Garden. Despite the prestige that it carried it was an ill-paid post with a salary of only £40 a year. George Don was recommended as Mackay's successor by Sir James Edward Smith and James Brodie of Brodie, the Lord Lieutenant of Nairnshire, both of whom had knowledge and experience of Doohillock and its owner. The recommendation was made to Professor Daniel Rutherford, the Regius Keeper of the Royal Garden. It was, in some ways, a surprising nomination, given the fact that Don had no formal qualifications, or indeed scientific training. What he did have was widespread recognition of his extensive practical experience and skills.

Don was at first reluctant to leave Forfar with the freedom it gave him for his expeditions, in exchange for the restricted city life that the new post would mean. With a wife and a growing family to be considered, the poor remuneration was also a disincentive. After some delay and persuasion, he finally accepted the offer and took up his duties at the end of 1802. He left the Doohillock garden in the capable hands of his father, Alexander Don, who in addition to his trade as a shoemaker, was also a keen and proficient gardener.

V   The Public Servant and Author

One silver lining for Don was the nearness of Dr Patrick Neill who lived in Canonmills, Edinburgh. Together they explored the district around Edinburgh, sharing their common interest in wild flowers and in cryptogams, those flowerless plants which reproduce through spores rather than seeds, and which include lichens, mosses and fungi. Altogether they discovered and noted almost a hundred of the former and two hundred of the latter which had never previously been recorded.

In recognition of his contribution to botany, Don was admitted as an Associate of the Linnean Society in 1803 in the first year of his Edinburgh appointment. This was a singular, but wholly appropriate honour for one who as a schoolboy twenty years earlier had devised his own system of classification for the plants in his relative's garden back in Menmuir.

The Linnean Society was founded in 1788 by Sir James Edward Smith, one of Don's two sponsors for the Superintendent's post at Holyrood. The Association was named after the Swedish naturalist Carl Linnaeus who did not invent, but who established in 1788, the so-called binomial classification of plants and animals which still remains in use. By and large a plant or animal was given two Latin or Latinised names, using one word to represent the genus and the other to represent the species. By tradition the generic name always begins with a capital letter. For example in the Latin name for the daisy, *Bellis perennis*, *Bellis* is the name of the genus to which the plant belongs and *perennis* distinguishes the species from others of the same genus. The Latin name is generally followed by the name or initial of the author who first described a particular species. For instance George Don discovered the wood stitchwort in Angus in 1813. The entry for this is *Stellaria nemorum* L. G. Don 1813.

As often as he could whilst he remained in Edinburgh, Don attended medical classes in the University. This may have been spurred on by his responsibility for the former physic garden at Holyrood House. Although he never qualified as a doctor, he was widely consulted after his return to Forfar. It was said of him that had he thrown up botany he would have done well in his new profession, for he was successful at first; but this business dwindled in consequence of his being continually out of the way when wanted, in search of new botanical discoveries – 'a pursuit ill-adapted for a poor man with a large family.' He was apparently sometimes approached for medical advice by post. In a letter to Mr David Booth, in Edinburgh, for example he wrote 'have sent the powder for Scrofula, with directions'.

The superintendence of the Royal Botanic Garden was a demanding job, but Don maintained his overwhelming passion for exploring the Scottish countryside. Apart from his co-operation with Dr Patrick Neill he carried on a frequent correspondence with the best botanists in England as well as in Scotland.

In 1804 he ventured into print through the good offices of Dr Neill. In that year he published the first edition of *Herbarium Britannicum*, which was dedicated by permission to Sir Joseph Banks, President of the Royal Society of London. Four folio fasciculi, or instalments, were to be published each year. Each of these containing twenty-five specimens of dried plants would be issued every year until a complete collection of nature dried plants had been made. This was a grand design and each edition was a rare work. One dried specimen of each plant was included in the *Herbarium* and enclosed in a small pocket was another specimen of the same plant that could be moistened for minute examination. The name and habitation were printed on a slip of paper and pasted on the page opposite. Only nine of the fasciculi were printed in Edinburgh between 1804 and 1806, although fasciculus 9 was not issued before 1812 or 1813. Some parts of the herbarium are now in the care of the Museum in Montrose. In his introduction to the work Don wrote: 'Since the editor first began his botanical excursions into the Highlands of Scotland in the year 1779, he is confident, as he hopes he may mention it without the imputation of vanity, that he has traversed more of the Caledonian Alps than any botanist has ever done.' He repeatedly ranged over the fine mountains of Angus-shire, which surround the great district of Clova, where no one on a similar pursuit ever preceded him. He also searched the vast range of mountains which stretch about sixty miles through the Knoydart in Inverness-shire – a region which had never before been examined by a botanical eye.

One survival of Don's work is the single remaining volume of his own private herbarium which contains almost three hundred specimens of monocotyledons and which is now in the Department of Botany at Oxford.

The *Herbarium Britannicum* was not Don's only venture into publishing. He contributed to a number of works, none of which produced much financial reward. The one exception was his 'Observations of the Indigenous Grasses of Britain' which won him twenty guineas as the prize essay in 1807 of the Highland Society of Scotland. His only other major published work was a contribution to the Agricultural Survey of the County of Angus by the Revd James Headrick in 1813. This was his 'Account of the Native Plants in the County of Forfar, and the Animals to be found there' and was printed as a substantial appendix to Headrick's work, running to more than fifty close-packed pages.

Don's time in Edinburgh did not last long. Two reasons have been suggested for his departure from the Royal Botanic Garden in 1806, less than four years after his appointment. One is the 'want of cordiality' between the Superintendent and the Regius Keeper, Professor David Rutherford. Dr Neill claimed that, as a botanist, Don greatly excelled the Professor 'who was an accomplished chemist, but had little turn for botany'. The other reason, and probably the more compelling one, was Don's longing to get back to Doohillock and to his freedom to roam at will on botanical expeditions from Forfar.

VI   Back at Doohillock

Back home again at Doohillock and in his native Angus, George Don settled in to his former pattern of life, continuing the care of the botanic garden and combining it with the commercial necessity of running a market garden. His over-riding passion and interest, however, continued to be in his expeditions of botanical discovery.

His powers of observation had been remarked upon when he was still a boy in Forfar Parish School. Those powers increased as the years went by. An outstanding example of this was contained in a letter which he wrote to Dr Patrick Neill which was later published in *Memoirs of the Caledonian Horticultural Society*. The subject of the letter was a matter that, at the time, was considerably exercising the minds of naturalists – the decline of the Scots fir. It was dated 11 February 1811.

> During the winter of 1810, when walking in some woods in the neighbourhood of Forfar, composed of Scotch fir, for the purpose of collecting mosses and lichens, my attention was drawn from these humble tribes by observing the very dissimilar appearances of different trees of what botanists consider as one species of *Pinus*, the *Pinus sylvestris*.
>
> After examining a great number of trees, I became satisfied that it is possible to distinguish in our plantations at least four varieties, and one of these indeed is of so fixed and marked a character that it may probably be entitled to rank as a species.
>
> It may here be proper to state that lately, while observing the cutting down of a fir plantation near Forfar, I was not a little surprised at the great difference in the size, and consequently in the value, of some of the trees in comparison with others of the species, the difference in value being not less than four times, and in some individual trees exceeding six times. I was at a loss to account for this fact, as the trees were growing promiscuously in the same soil and situation, and had been equally thinned. On more minutely examining those that far exceeded the others in size, I perceived that these were all of that variety which I have suggested as probably entitled to rank as a species.
>
> About a month ago, I re-examined the varieties of the *Pinus sylvestris*, in order to collect some cones of such variety for seed, and I was then led to examine them with considerable care.

Don then went on to describe the different varieties of the tree which he measured from one to four, the latter being the strongest and hardiest and, at one time, the most common fir tree in Scotland. This had been replaced by the variety he called number two which was both short-lived and stunted in appearance. He noted the quite distinct differences in leaves, in the disposition of branches, in the bark and in the appearance of the cones.

> I think this is the most natural way of accounting for the supposed decline of the Scots fir in this country, for two reasons – first, because variety four still retains all the good qualities ever ascribed to the Scots fir; second, as variety two produces its cones much more freely

8 *Forfar Loch*

than the other, the seed gatherers who were not paid by the quality but by the quantity would seize upon the former and reject the latter.

Don planted seeds of all four varieties at Doohillock and declared his intention of reporting upon the outcome. Sadly, he died before he was able to do this, but his early observations had defined the problem and its causes that were of so much concern to Scottish naturalists at the time.

From childhood George Don had taken a special interest in the plants of his immediate neighbourhood. This interest had borne fruit in his co-operation with Dr. Neill in Edinburgh where their exploration of the area around the capital city had produced extremely valuable additions to botanical knowledge. Now he was back on his own territory there was plenty of scope for exploration around the Royal Burgh. Amongst other places within easy walking distance from Doohillock were the banks of Forfar Loch, Restenneth Moss, Rescobie and the rocks on the Hill of Turin. All of these produced plants for the Botanic Garden.

His great love was for Clova. In his account of the plants and animals of Forfarshire appended to Headrick's *Agricultural Survey* of the county, Don wrote that 'not even Ben Nevis, Ben Lawers and Ben Lomond and the highest mountains of Cairngorm taken together can furnish such botanical treasures as are to be met with on the mountains of Clova'. In fact most of his discoveries came from 'Angus-shire' and especially from Clova. Many of them were confirmed by Thomas Drummond, gardener at Inverarity, near Forfar, who was to succeed Don at Doohillock in 1814. It was Drummond, incidentally, who made one of the earliest and most valuable contributions to botanical knowledge of the mosses of Scotland.

With motor travel in the twenty-first century, it is easy to forget the difficulties and hardships that would have been involved for Don's exploration of an area such as Clova two hundred years ago. It would at the least have taken five hours strenuous walking simply to get from Doohillock into Glen Clova proper without taking into account the further penetration needed to reach the head of the glen near Braedownie. From there two roadless valleys fork north-westward further into the high hills towards the White Mounth and Lochnagar rising to close on four thousand feet above sea level. A plant gathering expedition into this wild and lonely area not only called for careful observation and botanical expertise, but for considerable stamina as well.

Fortunately George Don was a strong and powerful man, but his preparations for these expeditions seem minimal to modern minds. He was often away in the hills for a week at a time, covering huge distances and sleeping rough. His normal equipment was a plaid for clothing and a bag of oatmeal, bread and cheese for food. For an extended journey he would carry two extra shirts, and when he felt a change was needed he would put one of these on top of the shirt he was already wearing. Occasionally he would be accompanied by a worker from Doohillock which cannot always have been a welcome prospect for someone who did not share Don's passion.

George Don's achievement in discovering such an astonishing collection of plants – and particularly alpines – probably owed much to the specialist equipment which he had devised for his expeditions. He had his own vasculum (specimen collection box) which can still be seen in the Meffan Institute in Forfar's West High Street, barely a stone's throw from his childhood home in Little Causeway. He also carried a staff which Dr Neill described as extending to fifteen feet in length. This was crowned with a straddle with which Don was able to hook down plants from what would otherwise have been inaccessible ledges and crevices.

His field of exploration extended far beyond his beloved Clova and Angus-shire. The introduction to *Herbarium Britannicum* referred to the vast tract of mountains 'which stretch sixty miles into the Knoydart in Inverness-shire'. His records show him, too, to have climbed many of the Scottish mountains, amongst them Ben Lawers, Ben Lomond, Ben Nevis, Ben Macdhui, Lochnagar and Schiehallion.

It is no wonder, then, that he frequently returned to Forfar immensely hungry and immensely tired. The Roll of Honour in Alan Reid's *The Royal Burgh of Forfar* (1902) recounts one story of such a return when Don had lost count of the days of the week. A story of his own is indicative of his engrossment in a favourite pursuit. One Sunday morning he arrived at a country manse haggard and unkempt but laden with specimens. His friend, the minister, who was just getting ready for the morning service, expressed his surprise at the visit. "Do you know it is Sunday?" he asked, when Don simply replied "Oh! It's Sunday is it? Well, I've fair lost count this week." (The manse at which he arrived was St Vigean's near Arbroath where the minister at that time was the Revd James Aitken). The account goes on to say that Don told Aitken that if he were able to wash his hands and face he would go to the kirk, too. Not surprisingly he had fallen fast asleep before he could fulfil his intention.

What is surprising is that in the midst of such a busy and exhausting life, George Don could engage in political discussions and work for the common good. In 1795 he was involved in the initiation of the Forfar Library which was later succeeded by the Forfar Free Library in 1871. A leather bound Album, dated 1795 and preserved in the present library, has a Preface stating that it is there 'to catch the fugitive idea as it passes and to give it a form which may reflect the merits or defects to the mind of others'. It is recorded in the first Minute Book that 'the old library is a rich store, particularly in History and Geography, and containing a valuable collection of correspondence and narratives founded amongst the echoes of the first French Revolution'.

9 *Commemorative plaque in Don Street, Forfar*

VII   Clova – The Land of the Wild Flowers

There can be no doubt as to the truth of George Don's claim that botany was his favourite science. Nor can there be any doubt that Clova was the favourite area in which he chose to exercise that science. He has the distinction of having brought the flora of Clova in particular, and of Scotland in general, to the attention of a wider botanical interest than ever before.

It is not difficult to see why Clova held such an attraction for him. In 1974 another Angus man, the late Duncan Fraser of Montrose, published his *Glen of the Rowan Trees*, the glen reference being to Glenesk. Fortunately the author also turned his attention to Clova, and it is clear that he was as attracted by the flora there as George Don had been two hundred years before. In a chapter entitled 'The Land of the Wild Flowers' there is a description of that part of Angus that catches the essence of Clova.

Almost all that follows in this section comes from Duncan Fraser's description, and is printed with the permission of his family.

Much more interesting than the history of Glen Clova, however, are its mountain slopes and the plants which grow there. It displays all the evidence that it was carved by an Ice Age glacier. For mile after mile the broad valley runs flat between precipitous hills, and down at the foot of the glen are the moraines that the glacier left behind. High on the hillsides too you can see the corries that were scooped out by hanging glaciers more than ten thousand years ago. But Clova's plants of today are more exciting than its glaciers of long ago. The only mountain in Britain more widely famous for its alpine flora is Ben Lawers overlooking Loch Tay.

It was upwards of two centuries ago that the wealth of rare plants in Clova was discovered by George Don, a Forfar man who enriched science and made himself bankrupt by his love of the plants around him. Among a host of rarities he found three that had until then been thought to grow nowhere in Britain, and two others were completely unknown even as foreign plants to the botanists of his day.

He made his rarest finds in almost inaccessible places, high among the hills, but it was not by accident that he found them there. Even at the foot of the glen, on the edge of the Lowlands, there were flowers that were little known. And all along the road to the hills they sign-posted the way to what lay beyond. By the time Don had covered the eleven miles up to the Milton he was in the heart of the flower country.

There, at a height of fully 2000 ft. above sea level, in one of those corries sculptured in the Ice Age, he found a gloomy loch almost encircled by precipices. It was on the banks of the Corrie Burn, close to this lochan, that Don made some of his finds. He made others beside Loch Wharrel, another hill loch a mile to the south-east of Brandy. High on the hills across the glen, just opposite the inn, you might have seen him joyfully on his hands and knees on the rocky bank of some waterfall near the summit of Carlowie, and higher still on the Bassies and the Scorie near Braedownie. The native Scottish azalea, the dwarf *A. procumbens*, was among the plants he found there, covering the ground with its splash of crimson flowers. In the corrie of Ben Hard, too, he saw for the first time the alpine *Saxifraga nivalis*. And snow saxifrage was a name which suited it. When you looked at its compact rosette of leaves, you got the feeling that Nature had designed it specially for survival, no matter how fiercely the blizzards might sweep across the high precipices where it made its home.

But it was beyond Braedownie, in Glen Doll and along the cliffs of Glen Fee, that this Forfar naturalist found the rarest of his plants. The blue alpine sow-thistle was one of these, growing on a rock ledge by a waterfall, high on Craig Maid. The yellow oxytropis *(O.campestris)* was another. It was to be seen in one place only, clinging all the way up a narrow vertical strip of rock, and spreading neither to left or right. It grew nowhere else in the glen. As far as the botanists of his day were aware, you could have searched all the rest of Britain without finding it. Farther on between Glen Doll and the top of Glen Isla he found something no less rare, the mountain sedge *Carex ruriflora* in a peat bog on Little Kilrannoch.

The night he found that sedge he probably pitched his tent among the lonely hills. He often slept out of doors. But other botanists who followed him to Little Kilrannoch in succeeding years, preferred the shelter of a roof for the night in the nearby shieling of Lunkar at the head of Glen Doll. The plant-hunters and the shepherds sat round the fire on a sunk, a bench made of turf, to eat their meal and they slept on a heather bed.

Maybe, when the botanists returned from the wilds of Glen Doll to their feather beds at home, they vaguely wondered whether Don had imagined his more startling discoveries. It was hard to keep up with an enthusiast like him. He had records of upwards of ninety local flowering plants, every one of them regarded elsewhere as extremely rare.

There was, for example, the alpine coltsfoot, well known in the Austrian mountains as *Homogyne alpina* but utterly unknown in Britain. In 1813, a year before he died, he claimed that he had just seen it with its kidney-shaped leaves and woolly stem, on rocks by a rivulet in the high mountains of Clova. He even had a specimen to prove it. But the years slipped past and no one else could find any trace of this alpine coltsfoot. So the certainty grew that he was wrong. Someone, obviously, had sent him a cutting from Austria for his garden and he had forgotten where it came from. That was the only logical conclusion – until another botanist rediscovered it in 1951. It covered only a few square feet of ground and it was almost, perhaps exactly, at the spot described by Don over a hundred and fifty years before.

But he was not the only field botanist who made finds that got lost again in Clova. A few years after his death the alpine milk-vetch with its purple clusters was seen by another botanist and half-a-century passed before it was re-discovered where it had been before. So there is still the possibility that two other plants described by Don – the *Potentilla tridentata* and the *Ranunculus alpestris* – will some day come to light there again. The find of individual rarities, however, was not his greatest achievement. It was Clova itself that he discovered and he blazed a trail which countless others have followed since.

Yet there were some things that he did not know about the plants on those hills. He was dead thirty years before anyone began to realise that there was something slightly odd about many of the rarest of them. They were more like plants of the Arctic than those you would find on a

Swiss mountainside. And that induced a couple of scientists to do a traffic census at the haunts of those Clova rarities, some eighty years ago. They made some remarkable discoveries while they carefully counted the bees and other insects that visited the flowers. They found that evolution was playing no part in the life of some of those plants. Unable to rely on cross-pollination, they had become self-pollinating and unchangeable. And yet by a miracle they had survived in their present locality, not just for a few years but for ten thousand years and more. They were as old at least as the glacier which once filled the bed of Glen Clova.

It has not been easy, of course, to keep them safe from the collectors. William Gardiner, in his *"Flora of Forfarshire"* in 1848, described the glen's unique *Carex rariflora*. Few botanists, he wrote, cared to leave the district without a sample of it. That same year the professor of botany at Edinburgh University wrote an article in a learned magazine about an excursion he made with a party of students to Clova. There was something peculiarly attractive about the collecting of alpine plants, he wrote, and at the head of the list of attractions he put their comparative rarity. Then he named the ones of special interest and among these he included the blue sow-thistle, the yellow oxytropis, the red mountain catchfly and the alpine milk-vetch. Very few collectors would have been needed to make those vanish from the British flora. It was not the collectors who saved them for posterity but the fact that often they thrived only on inaccessible ledges. At times, of course, even Nature herself was no help. By far the best place in Britain to see the blue sow-thistle was at one spot in Glen Clova – until a few years ago when the plant disappeared in a flood.

Yet some of the rare ones which Don knew so well are still surviving and no less rare today. The yellow oxytropis still grows in Glen Fee. The only other place in Britain where you will see it is in a remote part of the Perthshire Highlands. The red mountain catchfly still grows on Little Kilrannoch. There is only one other place for it as well, in a precipitous gully on the Cumberland hills. Loveliest of the lady's mantles is the alpine *Alchemilla conjunta*, a favourite in gardens up and down the county. In its native state you can see it growing in only two places – far up Glen Clova and in Glen Sannox on the Isle of Arran. *Carex norvegica*, the rarest of the sedges, was not found until after Don's death, but this too is known to grow in only one place outside of the Clova district, in Glen Lyon in Perthshire. The rarest of all those mountain plants is *Homogyne alpina*, which is not known to exist anywhere in Britain except in Clova.

There are others, scarcely less rare than these, like that "sweetest and loveliest of our native flowers", the two-flowered linnaea, which grows in Glen Doll. And any expert will tell you that the very rare horsetail *Equisetum pratense* and that beautiful woolly willow *Salex lanata* nowhere look quite as magnificent as in their stations in Glen Fee. Clova is a district which is special by any standard.

VIII   Brickbats and Bouquets

A physical description of the Botanic Garden at Doohillock appeared in Chapter III. In the *Scots Magazine* in 1809 a brief description was given of the plants growing there at that time. The Latin names of the different plants will be unfamiliar to the general reader who is not a botanist, but they are included in this extract to indicate the breadth of Don's collection.

> No place could be found more favourable for alpines and aquatics, which are in general found to be of rather difficult cultivation, but which flourish here as in their native habitats. For hardy herbaceous plants in general, the middle parts of the garden are well adapted. To give some idea of the extent of the collection, I shall mention the number of species of several genera which are at present actually growing in the garden. Of the genus *Veronica*, there are 55 species; of *Salvia*, 50 species; *Campanula*, 44; *Allium*, 40; *Saxifraga*, 46 – some of the rarest ones, as *S. coesia, petroea, rivulario*, etc.; *Dianthus*, about 20 species; *Cucubalus*, 13 – being the whole ever cultivated in Britain; Silene, nearly 50; *Fumaria*, 14; the genera *Onorsis, Lathyrus*, and *Vicia*, almost complete; *Astralagus*, 40 species; *Trifolium*, no fewer than 69; *Hieracium*, 44. It were needless to enumerate more. The botanist will form a due estimate of this collection on being only told that he may here see upwards of 60 species of *Carex* flourishing in great perfection. The agriculturist may here find the whole of the hardy *Gramina*, carefully distinguished and arranged, amounting to over 100 kinds.

> This season Mr. Don has introduced several hundred species of hardy plants, most of which we are told, have never before been cultivated in Scotland. Among the rare British plants at present in flower in this garden may be mentioned the elegant little grass called *Knappia agrostidea* (*Agrostus minima* of Dr. Smith); and the *Holosteum umbellatum*. Among the hardy exotics now in flower, the *Panaz quinquefolia* (the root of which constitutes the famous panacea of China called ginseng), is the most remarkable. There are certainly very few living specimens of this plant in Scotland; and we have not heard before of its flowering in this country. The *Balebarda fragaroides,* brought from North America to France by Michaux, and only lately imported into Britain, has already found its way into Mr. Don's collection. It is entirely a new plant, belonging to *Icosandria Polygynia*, and naturally allied to the Geums. The Forfar Garden, it must, however, in conclusion, be confessed, makes very little external show, being in a great measure destitute of the ornament which arises from neat alleys, with hedges or edgings, or well-laid-out or well-kept gravel walks. It is, in fact, merely an uncommonly excellent collection of hardy plants; and while it would doubtless fail to please the lover of tasteful gardening, it would as certainly prove highly interesting to the botanist, and to the curious cultivator. Mr. Don, we have been told, has an ample nursery of rare hardy plants, for which he receives orders from the curious in different parts of Britain; and, when the proceeds of these shall enable him, we understand it to be his intention to improve the exterior appearance of the garden.

Mr Don was by this time well known to all the leading naturalists of that day as a wonderful discoverer and collector and cultivator of rare plants. He could number among his correspondents such names as Dr Patrick Neill (who we know came to Forfar specially to see him); Sir Joseph Banks, President of the Royal Society of London; Sir James Smith, President of the Linnéan Society of London; and in Forfar itself Mr Headrick, the minister of Dunnichen, Mr Dempster, and Dr Jamieson were all great friends of his, and often visited him at his famous garden.

Don continued to maintain a remarkably large stock of plants at Doohillock, and must eventually have acquired a greenhouse. A catalogue of plants that he was offering in 1813 contains more than two thousand specimens and these included some greenhouse plants.

As well as alpine plants from Clova and other Highland areas there were specimens from the lochs and marshes around Forfar and a large number of flowering plants, mosses and lichens from the shores stretching from Montrose to Dundee and along the banks of the River Tay on the western side of the county. In his appendix to Headrick's *General Survey* of 1812, Don set out the various categories of plants. Although it was not an exhaustive list, he ended 'the larger plants (numbering over three hundred) contained in the above list may be seen in a growing state in my botanic gardens in Forfar where I have the most extensive collection of hardy plants in Scotland'. The Doohillock collection comprehensively represented the whole flora of Don's Angus-shire.

It was he who had pioneered the exploration of Clova, and it was to be another dozen years or so before botanists from elsewhere began to pay attention to the area.

Later in the nineteenth century some doubts were raised about the authenticity of Don's reporting and identification. The principal sources for these doubts were contained in two publications – *British Flora* by George Arnott and *Cybele Britannieum* by Hewett Watson. Fortunately, Don's reputation had a stalwart defender in the person of George Claridge Druce, a Fellow of the Royal Society, Fielding Curator in the department of Botany in Oxford and Secretary of the British Botanical Exchange Club. He was one of the foremost botanists in the later nineteenth and early twentieth centuries. In the *Scottish Naturalist* of 1883-1884 Druce, who was then in his early thirties, addressed with a forensic skill the various criticisms levelled by Arnott and Watson against Don.

Druce began by saying that having read the various paragraphs in the books by Arnott and Watson – 'to him, and probably to any other young botanist, Don seemed to rank first in the list of loose, if not unreliable, recorders. I pictured to myself some long-legged, red-headed Celt loaded with a wallet of garden plants going out surreptitiously to place one here, one there, later on recording with a flourish of trumpets the addition of a new plant to the British flora, and gaining by this detestable trick any amount of kudos and reception from the easily imposed upon botanists of their day'.

Both Arnott and Watson were excellent Botanists, but neither had spent days and nights in summer and winter in the more remote areas of Scotland's mountains and glens. Arnott's criticism of Don was the more vicious of the two. Of him, Druce wrote:

He evidently considered plants could be diagnosed with all the precision of a chemical formula, and never mastered the elementary idea of the extreme vulnerability of animate nature to change, so that split species or intermediate forms receive at his hands most cavalier treatment, or are summarily dismissed from notice as hybrids. Upon this arbitrary botanist Don exercised a most morbid influence, so that no suggestion of dishonesty or bad faith seemed strong enough to apply to him, for it is not only that he charged Don with making mistakes – mistakes that are most egregious blunders – but that he asserts that no credence can be given to Don's statements because he intentionally misled, either by making false records, or sowing plants, or by distributing plants from his garden as if they had been gathered wild.

Watson was less vituperative than Arnott and certainly did not impute bad faith to Don. 'I do not see that anything at present known of his (Don's) conduct or any necessary inferences from known facts would sufficiently warrant us in charging him with intentional deception or wilful falsehood.' In both authors there can be detected something of intellectual or scientific snobbery based on George Don's lack of a formal scientific education and the fact that he had to engage in the trade of his cultivated plant specimens. So Watson wrote:

> It appears that Don was in the habit of bringing the plants found on his excursions into his garden for cultivation, and there can scarcely be a doubt that he occasionally gave or sold plants from his garden without explaining that they were not sent direct from native localities, but indirectly through his own garden. When we add to this the fact that botanists were far less particular about the nativity of specimens half a century ago and also that George Don not having had a scientific education was loose in his indications and reports of localities, the presumption of frequent errors becomes very strong.

George Druce tackled each of the cases on which Arnott and Watson had cast doubt – in most instances demolishing the objections. He did, however, concede that Don had little knowledge of geographical distribution or of the microscopic living tissue of organisms, but what he did have was a fully developed naturalist's keenness of eye, a scientific love of classification which showed itself by minute differences.

> Above all Don had that divine fire (often absent from the endowed professor) which kept him at unwearied labour plodding, it may be, over the spongy morass or breasting the high and solitary moorland, or climbing with all a fowler's zeal up the high, rocky crags of that lovely district of Clova, itself not the least of his discoveries … Some of us who have trodden over the same lovely district, and who have gathered in the identical localities many of the rich treasures Don has made known, can feel some gratitude for his labour and respect for his memory and life.

Druce's gratitude was an echo of that of numerous previous botanists of great distinction, including Sir James Smith, founder of the Linnean Society; Sir Joseph Banks, President of the

Royal Society who had accompanied Captain Cook on the Endeavour around the world; and John Knapp the publisher of *Gramma Britannica* or *Representation of the British Grasses*.

Druce concluded his defence by setting out twenty-two of the most important of Don's undisputed discoveries and more than fifty other species (not including lichens and mosses) as British plants, of which he was amongst the earliest recorders, or where he had added new localities to those previously known for them.

10 *Ben Lawers*

IX   The Closing Years

George Don was forty-three years old when he returned to Forfar from his post as Superintendent of the Royal Botanic Garden in Edinburgh. He was to have only six years back at Doohillock before his death at the beginning of 1814. It would be trite, but true, to say that they were action-packed years. His passion for plant-collecting continued unabated. It was a passion that both made him, and un-made him.

His attention to what he described as 'his favourite science' prevented him from giving proper attention to the nursery, and in particular to that part of it which served as a market garden. It was that section of Doohillock which helped to provide a day-to-day income and which needed day-to-day attention. From time to time he was given a commission to make up an herbarium, and on one occasion to form a complete herbarium of all known British plants. But these were not occupations that provided a regular income for a man with a large family. It was not that Don or his wife were extravagant – all the indications are that they lived extremely frugally.

His financial affairs grew more and more difficult. So much so that in 1812 he was forced to enter into some kind of arrangement with his creditors, which at least prevented Doohillock from closure. This did not deter Don from his botanical excursions in search of plants that

11 *Alpine forget-me-not*

hitherto had never been recorded. In the same year as he signed his Deed of Arrangement he discovered a new member of the Pea family *(Oxytropus compestris)* on a rock in Glen Doll in Clova. Nevertheless, it would appear that he never really recovered from losing his independence in the face of his creditors.

The following year the family had become so poor that they were reported to be reduced to dependence for their daily bread upon the charity of neighbours. Towards the end of the autumn of 1813, Don came home from one of his excursions suffering from a severe cold. He ignored this for some time and continued working. Eventually he was compelled to take to his bed. He developed what was at first a sore, and then a suppurating, throat and for six weeks suffered excruciating agony before his death on 14 January 1814. He was only forty-nine years old.

Tradition has it that his funeral was attended by the largest number of mourners that the Burgh had ever witnessed up to that time. He was buried in an unmarked grave quite close to the east door of Forfar Old Parish Church.

In *Seeds of Blood and Beauty* Ann Lindsay paints a graphic picture of Don's eldest son shuffling miserably in the funeral procession as it followed his father's coffin up the East High Street to the burial ground of the Old Parish Church. 'The outlook was indeed bleak for him … for George's widow it must have been like staring into a bottomless pit'.

Her husband's death had left the family destitute. Caroline had five sons and a daughter to provide for – George was approaching his sixteenth birthday and was already committed to the Chelsea Garden. The next son, David, was an apprentice with Dickson's of Broughton in Leith Walk, Edinburgh, and was only fourteen years old at the time.

Both brothers had spent a considerable time working in the Doohillock Garden. George, in particular, was bitterly resentful at the way in which the creditors had treated his father, as a consequence of which the whole family had been forced into dire poverty. It was a resentment that lasted until he was well into middle life.

The immediate need was to help the family to survive. The two oldest boys were not yet sufficiently knowledgeable to manage the Botanic Garden, and in any event that would have meant frustrating their desire to become botanists like their father. Three particular friends of George Don engaged themselves in trying to raise funds to help. They were William Roberts and John Rodger, both Forfar residents, and David Booth, a lexicographer from Newburgh in Fife. Together they wrote to a number of Don's correspondents including the President and Vice-President of the Linnean Society and the Bishop of Carlisle.

Booth wrote to Sir John Smith:

> Mr. Don died in extreme poverty, having been obliged during his illness to accept the private donations of friendship, which must have ill accorded with his independence of mind. He has left a widow and six children, four of whom are incapable of labour. Two sons (who, I suppose, are from fifteen to sixteen years of age) have been accustomed to work in the garden, but they are quite unfit either to continue or to sell off the valuable collection of plants which it contains. Indeed, there is no-one in that quarter that can appreciate their value, and

what has been collected by the labour of years, will most probably be thrown out as useless cumberers of the ground.

The appeal was reasonably successful and afforded Don's widow time to sell the plants and arrange for the transfer of the tenancy. A letter from one of Don's grandsons, George Alexander Don, head-gardener to the Right Honourable Beresford Hope M.P. in Bedgebury, Kent, completes the story of Doohillock:

> After his death in 1814, my grandmother sold all the nursery stock and went to live in Newburgh, Fife. As my grandfather left no provision for wife or family, she had a hard struggle to bring up and educate the family decently. Out of the large family of fifteen she bore to my grandfather, only six reached adult age, and one of them, the eldest and the only girl, I think, died soon after her father and before they left Forfar. All the five sons were bred to gardening … The two eldest sons, David and George, did not long follow gardening, but having ability and a great love for botanical science, struck out a higher and more congenial path for themselves. The others were not so fortunate, although they all held good appointments in their calling.

The nursery stock was sold to Thomas Drummond, a son of the head gardener on the Fothringham Estate at Inverarity. Caroline also transferred the tenancy to him. Drummond at twenty, was only five years older than George Don junior, who believed he was as capable of taking on and running the nursery as Drummond would be. This only served to compound his resentment at the way in which the family had been treated. Dr. Patrick Neill counselled the young Don against taking on such a responsibility, advice which resulted in a coolness between the two which was only restored when Don eventually acknowledged that Dr Neill had been right.

Drummond continued to maintain the garden for another ten or twelve years. When he left, the garden was parcelled out. The tenant of the house that Don built less than twenty years earlier, was given part of the garden, and the rest let out as private gardens. Soon afterwards the house became a public-house by which time there were no traces of the former glory of Doohillock.

The hillock itself and the ground to the west of it became part of a linen works and today is the site of a residential home. The lake, so carefully stocked by Don with aquatic plants and fish, was filled in, levelled and covered with grass.

Doohillock was gone.

Thomas Drummond went on to play a distinguished part in the botanical world. He left Doohillock in 1825 to become assistant naturalist in Captain Sir John Franklin's second Arctic expedition. For two years he made collections in the regions of British Columbia, Alberta, Saskatchewan and Manitoba. It was an epic journey and a highly successful one. On his return he became curator of the Belfast Botanic Garden before setting out on his second overseas journey in 1831 to New Orleans and Texas. He died in Havana in 1835 at the age of forty-eight.

Meanwhile, George Don's grave lay unmarked in the burial ground of Forfar Old Parish Church. In fact, it was to be almost a century – ninety-six years after his death – before a suitable memorial was erected to mark the last resting-place of this distinguished son of Forfar.

His memory was preserved, however, by one of the Angus bards – James Ross of Forfar – in the lines of this elegy:

> Has Forfar now no melancholy muse
> To moan for genius which so brightly shone?
> If such there be, let her with tearful dews
> Moisten the turf that marks the grave of DON.
>
> Bid roses rise, and violets round it bloom;
> Bid sweetbriars breathe their balm in churchyard lone;
> Bid bays embowering form a verdant tomb,
> To memorise thy native genius, DON.
>
> Bid every plant that decks the desert droop,
> And all the grassy tribe on swamps lie prone;
> And lowly bend bid Dovehill's flowery group –
> For Death has chilled the fostering hand of DON.
>
> Bid sylvan songsters that in shades sojourn
> Begin his elegy with plaintive tone;
> Bid Spring, the flowery-mantled maiden, mourn,
> For many a vernal debt is due to DON.
>
> Bid blackbirds sorrowing sing upon the spray,
> For he who nursed the nestling-bush is gone;
> And linnet tune a mournful matin lay,
> Till twilight thrush trill tenderly for DON.
>
> With solemn sound bid Naiads torrents pour,
> Till echoing woods and rocks responsive moan;
> And Dryads, from their summer-incensed bower,
> Lamenting, join the dirge that peals for DON.
>
> No more will he the rural scene survey,
> And deck like Darwin the Linnean throne;
> Then let some lyre with laurel-lasting lay
> Enbalm in elegy the name of DON.

X   After Doohillock

No account of the life of George Don would be complete without a record of the influence he must have had, consciously or unconsciously, upon his two eldest sons. Left fatherless as teenagers, they had already caught enough of their father's passion for his 'favourite science' to have encouraged them both to become botanists.

Writing on 18 January 1814, four days after Don's death, William Roberts (a friend of Don) in a letter to David Booth said of the two boys:

> The elder sons, George and David, have studied botany under their father, and have made considerable proficiency. They know the greater part of the immense variety of plants in his botanic garden. The second son, David, is a fine boy of about sixteen years of age, modest, communicative and sensible, and the knowledge he has already acquired of plants would astonish you. Were his genius to be properly cultivated, I have not a doubt that he could be little inferior to what his father was in the service of botany. Perhaps a small fund may be raised to preserve the family from starving, and to enable the two elder sons to follow their pursuit in the knowledge of botany, and if they meet with any encouragement, they may possibly become an acquisition to the world. They are young and vigorous and able to traverse Alpine regions in the pursuit of plants.

Roberts' words were prophetic. Both George and David were enabled to follow their chosen profession, and both added distinction to their father's memory. Caroline Don had moved to Newburgh after her husband's death. The money that had been raised to help the family was being administered by David Booth, the recipient of Roberts' letter. Booth was himself resident in Newburgh and it may well have been through his good offices that the boys' education was progressed.

George had already begun work in the Chelsea Garden when his father died. In November 1821 he was sent to Brazil, the West Indies and Sierra Leone by the Royal Horticultural Society to collect plants. He sailed in the Iphigenia under Captain Edward Sabine, and his discoveries were later published in the *Transactions* of that Society. He followed in his father's footsteps by becoming an Associate of the Linnean Society in 1822 and nine years later was made a Fellow. He had a number of publications to his credit. His principal work was *A General System of Gardening and Botany* published in four volumes between 1832 and 1838, which is still regarded as a standard work. He died in London in 1857.

David Don was an apprentice in the gardens of Dicksons of Broughton in Edinburgh at the time of his father's death. In 1819 he joined his brother in London, working successively in the Chelsea garden of the Apothecaries Company, and then as Keeper of the Library and Herbarium of Aylmer Lambert, a leading botanist and plant collector. In 1822 he was appointed Librarian of the Linnean Society of which, like his brother and father before him, he became an Associate and then a Fellow. In 1836 he was appointed Professor of Botany at King's College, London, a post he held until his untimely death at the age of forty in the year 1840.

In a paper to the Perthshire Society of Natural Science in 1881 Mr John Knox, Headmaster of Forfar Burgh South School, lamented the fact that no memorial marked the grave of George Don.

> During his short life he did more than any other individual had ever done in stimulating the study of the botany of his native country and especially of the Highlands. During that time little or nothing had been done, in a comprehensive way at least, in the study and exposition of the native plants … The man is now little more than a shadow. Yet surely something can be done for past neglect. Will not the votaries of the science he loved so well and served so devotedly subscribe a few pounds for a simple stone to mark where his ashes rest?

Another twenty years were to pass before Knox's plea was to bear fruit. In his *Royal Burgh of Forfar*, published in 1902, Alan Reid noted that the name and fame of Don had quite recently been the subject of public attention. The desirability of a public memorial to his genius has been debated in *The Scotsman'*. An eloquent writer gave expression to his opinion in these stirring words:

> When a man, as Don did, extends the bounds of knowledge of his time, he places society under a debt which is not fully repaid by attending his funeral in large crowds … Ought society to do nothing for such men if by the defect of their qualities they are not practical?

The move to erect a suitable memorial gained momentum, largely due to the Forfar Field Club and to the unremitting efforts of Knox. Their labours were rewarded in 1910 when a memorial obelisk, seventeen feet in height and made of blue Aberdeen marble was erected over the grave of George Don. A plaque on the plinth at the foot of the obelisk reads: 'This monument was erected by public subscription through the efforts of the Forfar Field Club; John Knox, Schoolmaster in Forfar; and also of George Claridge Druce M.A.(Oxon)F.L.S., by whom it was unveiled, 8th September 1810.'

It was wholly fitting that it should have been George Druce who unveiled the memorial. For it was he who had sprung to the defence of Don's integrity a quarter of a century earlier, against the assaults of George Arnott and Hewett Watson. In introducing Druce, Knox said that if there was anyone in Britain naturally drawn to speak about Don, that person was George Druce. There was no one present who knew Scotland so well or who had tramped over mountains and moors in his botanical research as Dr Druce had done.

The inscription on the memorial is as under. It had taken a long time but for many people that day, it must have felt that a reproach had been removed.

> TO THE MEMORY OF GEORGE DON, BOTANIST, NATIVE OF FORFARSHIRE, WHO, AFTER A RESIDENCE OF MORE THAN TWENTY YEARS IN FORFAR, DIED AT THE DOO HILLOCK THERE, 15TH JANUARY, 1814.

DON WAS A MAN OF GENIUS, WHO, WITH FEW EDUCATIONAL ADVANTAGES, RAISED HIMSELF TO A HIGH PLACE IN THE RANKS OF THE BOTANISTS OF HIS DAY. HE ESTABLISHED A BOTANIC GARDEN IN FORFAR, WHICH CONTAINED A MOST EXTENSIVE COLLECTION OF NATIVE BRITISH PLANTS.

HE WAS SUPERINTENDENT FROM 1802 TO 1806 OF THE EDINBURGH ROYAL BOTANIC GARDENS. FOR HIS SERVICES TO BRITISH BOTANY HE WAS ELECTED IN 1803 AN ASSOCIATE OF THE LINNEAN SOCIETY OF LONDON.

IN 1813 HE PUBLISHED AN ACCOUNT OF THE NATIVE PLANTS OF THE COUNTY OF FORFAR. IN THE COURSE OF MANY JOURNEYS OF EXPLORATION MADE IN THE HIGHLANDS UNDER HARDSHIPS AND PRIVATIONS, HE ADDED LARGELY TO THE THEN EXISTING KNOWLEDGE OF THE FLORA OF HIS NATIVE COUNTRY; AND SO LONG AS THERE ARE STUDENTS OF THE ALPINE FLORA OF BRITAIN, HIS NAME WILL BE HELD IN AFFECTIONATE REMEMBRANCE. FIVE PLANTS PERPETUATE HIS NAME – A ROSE, ROSA DONIANA; A WILLOW, SALIX DONIANA; A GRASS, AGROPYRUM DONIANUM; AND TWO MOSSES, GRIMMIA DONIANA AND SALIGERIA DONIANA.

12 *Memorial obelisk – Forfar Old Churchyard*

13 *Rock sedge*

XI   Looking Back

The most striking paradox in the story of George Don is the contrast between the wealth of the legacy which he left to the world of scientific knowledge and the abject poverty in which he died. This was an aspect of his life which Dr George Druce underlined when he unveiled the marble obelisk in September 1910.

> As Walter Scott made known the beauties of Caledonia stern and wild, so Don in his narrower sphere laid open to the botanist of succeeding times the rich mountain flora which is so richly bestowed yet so often skilfully concealed in the corries and crags of the high hills, and this has created in the minds of us who follow the same pursuits a sense of gratitude and keen admiration for his toilsome and exacting and unremunerated labours. To those whose only standard of success is opulence his life would be pronounced a failure; to those who love ease and luxury his career would be looked upon as insanely miserable; yet I doubt if the wealthiest millionaire ever derived as much satisfaction from the accumulation of his riches as Don experienced in finding a new species ... Because of his independence of spirit and as a testimony to the example of patient and continued work at the Botany of his native land for which he did so much, we today in this ancient Burgh unveil this monument to his memory and give thanks to God for the life and labours of George Don.

Don's legacy was not confined to his great collection and classification of Scottish plants, but spilled over into the inspiration that his life's work gave to succeeding generations of collectors. Some of these continued their explorations in Scotland, but others went out to the four corners of the earth in their search. The names of John Mackay, Robert Brown and Patrick Neill have already been mentioned, but there were countless others who caught something of George Don's passion for enlarging the boundaries of botanical knowledge.

The most obvious influence was upon his own immediate family, and in particular upon his five surviving sons. The three youngest boys all became professional gardeners, and all eventually found well-paid positions south of the Border. A brief note of the subsequent careers of the two eldest sons – George and David – was set out in Chapter X. It says much for the training that George, junior had received at Doohillock that he was made foreman of The Chelsea Physic Garden of the Worshipful Company of Apothecaries at the age of seventeen. His brother David's abilities were equally apparent whilst he was still in his teens.

The memorial obelisk in Forfar Old churchyard refers to five plants which serve to perpetuate the name of George Don – a rose, a willow, a grass and two mosses. Those five are, of course, only a fraction of his discoveries. In the *Scottish Naturalist* of 1886 Dr. George Druce set out a definitive list of the more important of Don's undisputed discoveries, together with some fifty other species of which he had been amongst the earliest recorders.

A random glance at the undisputed list gives an idea of both the wide variety of plants which he discovered and also of the rich source that Clova had proved for Don in his collecting.

Alpine Catchfly (*Lychnis alpina*) — Clova

Rose (*Rosa Doniana*) — Forfar

Saxifrage (*Saxifraga platypetala*) — Clova

Alpine lettuce (*Lactuca alpina*)

Willow (*Salix lanata*) — Clova

Sedge (*Carex rariflora*) — Clova

Forget-me-not (*Myosotis repens*)

Wild oat (*Avena alpina*) — Clova

Foxtail (*Alopecurus alpinus*) — Lochnagar

In the face of reports that each county in the United Kingdom was losing an average of one wild flower species every year, Plantlife International launched a County Flowers Campaign in 2002. The Campaign invited members of the public to nominate and vote for a wild flower that would be properly representative of their county. There was a phenomenal response to the invitation. Hundreds of different wild flowers were put forward and votes arrived in their tens of thousands.

It was wholly appropriate that the successful nomination for the County of Angus – or Forfarshire – should have been one of George Don's earliest discoveries – found in Clova in 1795. Even today, almost the entire British population of this pretty pink-flowered alpine is still confined to a single remote hill-top in Angus, home to the Alpine Catchfly (*Lychnis alpina*).

Appendix

Specimens from the Forfar Museum Herbarium, entitled George Don material

Accession Number	Herbarium no.	Scientific name	English Name	Date	Location	E. H. Robertson notes
NH1982.3204	155	*Raphanus maritimus (Sm.)*	Sea Radish	1793	Side of Gairloch	
NH1982.3205	232	*Arenaria tenuifolia L.*	Fine-leaved Sandwort		Knaresborough	*Minuarta hybrida (Vill.) Schischk*
NH1982.3206	238b	*Arenaria fastiagata (Sm.)*	Sandwort	1800	Rocks, Clova	This discovery has never been confirmed.
NH1982.3207	373	*Astragalus glycyphyllos, L.*	Wild Liquorice	1800	Arbroath, seaside	
NH1982.3208	388	*Vicea lutea L.*	Yellow-vetch	6/1804	Bank East of North Queensferry	
NH1982.3209	625b	*Apuim repens (Jacq.) Reichb.f*	Creeping Marshwort	1800	Goalan Links [=Gullane]	And by him named.
NH1982.3210	738	*Filago minima (Sm.) Pers.*	Small Cudweed	16/6/89;1791	Reading District	
NH1982.3211	1525	*Eleocharis multicaulis (Sm.) Desv.*	Many-stalked Spike-Rush	1800	River Dee near Aberdeen. Also plentiful on the Pentland Hills	
NH1982.3212	1542	*Eriophorum alpinum L.*	Hair-tail Cotton Grass	7/1791	Restennet Moss. "I discovered this plant in July 1791, in the moss of Restennet near Forfar but it is now totally extirpated from thence by digging of marl and peats and at present we know of no British habitat for the plant."	*Scirpus hudsonianus (Michx.) Fernald.* Given to me by Mr. John Knox 1892. Extinct by 1804 due to the digging of marl and peats.

Norman Atkinson, 2007

14 *Cairn Inks or Bassies, Clova*

15 *Lychnis Alpina*

16 *Craig Mellon, Glen Doll*

17 *Towards Glen Doll*

18 *Dean Water,*
Forfar Loch Country Park